YOUR KNOWLEDGE HAS VALUE

- We will publish your bachelor's and master's thesis, essays and papers
- Your own eBook and book - sold worldwide in all relevant shops
- Earn money with each sale

Upload your text at www.GRIN.com and publish for free

Mutagenicity Assessment of Textile Dyes by Using Ames Test

Harpreet Kaur

Bibliographic information published by the German National Library:

The German National Library lists this publication in the National Bibliography; detailed bibliographic data are available on the Internet at http://dnb.dnb.de.

ISBN: 9783346494405
This book is also available as an ebook.

© GRIN Publishing GmbH
Nymphenburger Straße 86
80636 München

All rights reserved

Print and binding: Books on Demand GmbH, Norderstedt, Germany
Printed on acid-free paper from responsible sources.

The present work has been carefully prepared. Nevertheless, authors and publishers do not incur liability for the correctness of information, notes, links and advice as well as any printing errors.

GRIN web shop: https://www.grin.com/document/1060196

Mutagenicity assessment of disperse and Vat dyes by Ames test

Submitted in Partial Fulfillment of

The Requirement for the Award of Degree of

Master of Science in Microbiology

Submitted By: Harpreet kaur

Discipline of Biotechnology and Biosciences Lovely Professional University Phagwara, Punjab

ACKNOWLEGDEMENT

I would first like to thanks my thesis advisor **Dr. Joginder Singh Panwar, Associate Professor, School of Biosciences, Lovely Professional University** for his wise and painstaking guidance that he has given throughout this research work. His generous co-operation, suggestions, Constant encouragement and kind supervision, made this dissertation possible.

I express my gratitude to **Dr. Neeta Raj Sharma, Professor and Head of School of Biosciences, Lovely Professional University** their kind support and encouragement for my Research endeavor. I am highly indebted to **Deepika Ma'am** for her frequent help, care and Encouragement during my study period of completing this thesis. I would also like to express my deep thanks to my friends especially (Ayan Das) whose selfless cooperation helped me in my thesis. I am also thankful to staff members of library and all other teaching and non-teaching staff especially (John Masih) senior techncian of Lovely School of Biosciences, Lovely Professional University, for their kind care and cooperation.

Finally, I respectfully offer thanks to my beloved parents, especially my mom (Kamaljit kaur) For their love, mellifluous affection and sincerity which hearten me to achieve success in every Sphere of my life. I also pay thanks to my beloved sister (Jaspreet kaur), and to my husband (Kirandeep Singh) regarding their love, moral support and constant encouragement. Without their Encouragement and support, the present study would have been a mere dream.

Dated:

TABLE OF CONTENTS

Chapter: 1

Introduction

"The earth provides enough to satisfy every person's need, but not every person's greed"

-Mahatma Gandhi

The human life fitting to the natural variety of the earth's system and climate. The earth consists of plants, animals, human being, air, water, soil and land. Some conditions are necessary for the development of our life on earth. If these condition are present in unlimited from then it cause some problem 'As we know excess of everything is bad'. Activities done by human into natural environment are so widespread and silence in their consequences that they too affect the earth.

The researches done by scientists are also paly important role in protecting theEnvironment. Researchers tell the advantages of environment to the humans health economic, social and aesthetic harm. Result for this whole is that we are now aware that how we protect our environment from these natural hazardous things.

The environment which is free from these contaminations i.e. the environment which is safe and hygienic should provide us clean water, soil, and natural environment. In the past, men rely on inexhaustible resources which are provided by environment to us without a limit and the law of nature directed our evolution. But now in present world we started supplements on muscle power with exhaustible resources, coal, oil, and uranium. As a result, in some aspect We are directed our own evolution.

Environment pollution

Nature is a gift of God. We have no any right to destroy its beauty. We should not destroy its beauty and if we do so, we will have to face the consequences. We should take care of nature, because it is a basic need of our life but now pollution is the major problem on earth. Pollution is defined as the introduction of contaminants into the environment that results adverse change in environment. It is one of the major issues in the modern world. Pollution can be present in environment in the form of chemical substances or energy such as water, heat and light. Many Countries in the world having a specific connection with the effects of contaminated drinking water because of that water-borne diseases are a more common or increasing in the environment that lead to the major cause of morbidity and mortality (Clasen et al. 2007; WHO 2010). Contamination free drinking water is great significance for overall health and have considerable importance in young and infant child health and their survival (Anderson et al. 2002; Fewtrell et al. 2005; Ross et al. 1988; Vidysagar 2007). The World Health Organization (2005) estimates that worldwide about 1.8 million people die from diarrheal diseases. Persons with weak immune systems, such as those peoples who was suffering from AIDS are especially harmed to water- borne infections, even those peoples which are not typicallys howing an intention to cause bodily harm to healthy individuals (Kgalushi et al. 2004; Laurent, 2005)

Types of pollution

Air pollution

Release of chemicals into natural environment cause air pollution. For example chemicals like carbon monoxide, chlorofluorocarbon, and nitrogen oxide which is produce by industry and motor vehicles into the natural environment.

Light pollution

In light pollution light trespass, over illumination and astronomical interference is a major cause of pollution into the environment.

Noise pollution

Noise pollution includes aircraft noise, industrial noise, roadways motor vehicles noise which makes peoples sick and tensed. Also due to noise people brain not able to take rest.

Soil pollution

Soil pollution occur when soil are exposed to heavy metals, hydrocarbon, M TBE herbicides, pesticides, or chlorinated hydrocarbons.

Water pollution

Contamination which occurs in water bodies such as lakes, rivers, underground water etc. is known as water pollution. In this we mainly consider the contamination which is occur due to textile dyes. It occurs directly and indirectly into the water.

Pollution caused by textile industries

Dyes are mainly used to produce consumer's products like paints, textile printing inks, paper, and plastics. They are very complex and sensitive chemicals. They add color and pattern to the material. Many different vat dyes and dispersed dyes were used in the textile industries.

Textile dyes industries causes wide range of pollution in the world. The World Bank roughly calculated about 20% of the global industrial water pollution originates from the treatment and dyeing process in the textile industries. Textile production of dyes predicted to release aromatic amines (Benzedrine and toluidine), ammonia, alkali salts, toxic solids and large amount of pigments, chlorine, a known carcinogen. The dyes which are not treated cause chemical and biological changes in our aquatic system and human life. The presences of this compound also make water unhealthy and dangerous to human life. The polluted water also affects the soil health and agriculture field by contaminating its effect the germination and growth of crops. The textile industry in Ludhiana city in Punjab produces high water pollution which is dangerous to peoples in the region. Textile Dyes Synthetic dyes are majorly used in textile industries for dyeing, paper printing, color photography, pharmaceutical, food, cosmetics and other industries (Rafi, Franklin and Cerniglia,1990). It was estimated that about, 10,000 different dyes and pigments are used industrially, and over $7x10^5$tons of synthetic dyes are produced every year worldwide. In 1991, the world production of dyes was find approximately about 6, 68, 000 tones out of which azo dyes contribute about 70 % (ETAD, 1997). During dying process huge amount of azo dye is lost in wastewater.

Types of dyes

The first invented man made organic dye was, maven, which is discovered by scientist William Henry Perkin in 1856. Thousands of synthetic dyes have since been prepared, synthetic dyes can being replaced by natural dyes. Classification of dyes given below:

1. **Acidic dye**

   These are the dyes which are applicable to the fibers such as silk, wool, and
   these acidic dyes are soluble in water.

2. **Basic dyes**

   These are mainly applicable to a acrylic fibers for wool and silk. These dyes are water soluble cationic dyes. These dyes are also used for the coloration of paper.

3. **Direct or substantive dyes**

   By the addition of sodium chloride or sodium sulphate direct dyes are normally carried out at slightly or neutral dye bath. These dyes are used on cotton, paper, leather, wool, silk and nylon etc. These dyes are also used as PH indicators.

4. **Mordant dyes**

   The most important mordant dyes are synthetic mordant dyes or chrome dyes. They applied to wool. These mordant dyes contain heavy metals that are dangerous to health and extreme care has been taken when we use them.

5. **Vat dyes**

   Vat dyes are water insoluble and not able to dying the fiber directly. However, alkaline liquor reduction produces a water soluble alkali metal salt of the dye.

6. **Reactive dyes**

   These are the dyes which utilized the attached chromosphere to substituents that are capable of directly reacting with fibers substrate. The colvant bonds that attaches the reactive dye to normal fibers to make them most permanent dyes..

7. **Dispersed dyes**

Dispersed dye is water insoluble dye and also used to dyeing cellulose acetate and they can also use to dye nylon, cellulose, triestate and acrylic fibers.

8. **Other important dyes**

There are number of other classes of dyes including oxidation bases, laser dyes, and leather dyes, solvent dyes, carbine dyes, carbine dyes.

The chemicals which are used to produce dyes today are highly carcinogenic, or even explosive. Aniline is the popular group of dyes known as azo dyes which are extremely poisons and dangerous to handle and work with, also being highly burnable. Other harmful chemicals which are used in the dying process include:

1. Dioxin – a carcinogen and possible hormone disrupter;
2. Toxic heavy metals such as chrome, copper and zinc-known carcinogens; and
3. Formaldehyde, a suspected carcinogen.

Disperse dye almost totally insoluble in water it exists as a dispersion of microscopic particles, with only small amount in true solution. Disperse dye extensively used in textile industries for dyeing. Azo dyes which are widely used in textile industries which are one of the largest class of synthetic organic dyes and that having nitrogen as the azo group. Azo dye has been shown to destroy natural ecosystem when these azo dyes was discharged into water system. At the same time, the more drastic dyeing conditions and improved methods which developed along with the newer fibers have allowed the use of more complex molecular structures, although many acetate dyes are still used on polyesters and nylon. Azo, anthraquinone, o-nitro diphenylamine, methine and quinophthalone chemical classes provide the major commercial disperse dyes. None of these is a new chromophoric system, but intensive investigation and novel elegant synthetic methods have resulted in dyes exhibiting superior properties, especially on the newer fibers. Disperse dyes are also used in the newly popular transfer print process, which requires coloring matters of poor sublimation fastness, a source of trouble in conventional printing.

The accurate amount of dye produced in the world is not known. Exact data on the quantity of dyes discharge in the environment are also not available. Globally, accessing to freshwater is becoming more acute every day. Studies have shown that many microorganism catabolise these synthetic dyes as sole source of carbon and energy. Instead of dispersed or azo dyes there are also many other dyes which causes water contamination. These all dyes contaminants the water and through water it can enter into human body and cause mutation which results into various effect to human health it can also be carcinogenic that is cause cancer in human.

.Studies have also shown that many microorganisms catabolise these synthetic dyes as sole source of carbon and energy. It was shown that some bacteria can also be used to degrade complex Azo compounds like 1-(4'-carbophenylazo)-4-naphthol. Hence, various type of dyes which are use at large scale in industries containing carcinogenic substances which when thrown into water cause contamination of water that lead to cause various disease in human. Different dyes which we use in textile industries are:

Hence, dyes which used in textile industries are more harmful for the people who live near the area. Microorganism growing in the industries effluents containing various dyes are cultured and studied for their potential in degrading those harmful dyes. These microbes basically use these dyes as a source of carbon, nitrogen, energy

CHAPTER: 2

AIM, SCOPE AND OBJECTIVE OF THE WORK

The research work aim was to assess the mutagenicity or carcinogenicity of textile dyes to the people live near to the textile mills.

Objective

- Survey of Ludhiana textile mills for knowing different disperse and vat dyes used throughout the year
- To perform an assay to determine the mutagenicity of disperse and vat dyes (Ames test)
- Data Analysis for Ames test

SCOPE OF THE WORK

Textile dyes were used in colouring the clothes.Textile wastewater includes various types of dyesand chemicals that make the environmental challenge for textile industry. Main pollutants in textile wastewater came from dyes.

Textile industry effluent contains these toxic dyes. These toxic compounds enter into the surface water contaminate the surface water and it is used for irrigation and drinking purposes. Also, farmers use water from the rivers for agricultural purposes and the nearby areas peoples of the town, use both the surface and underground water as potable water, it is very harmful and unsafe to discharge this effluent into water body.

Here, present work is focused on the characterization of various disperse and vat dyes used by textile industries of Ludhiana for mutagenicity assessment.

CHAPTER: 3

Review of literature

Micro-organisms find to have much credit that makes them interesting for use in screening of effluents and chemicals for their toxicity. Approximately more than 200 tests preform on micro-organisms, insects, plants and animals, have been developed over the last 20-25 years, for the identification of chemical agents that cause genetic hazard to humans (Waters et al., 1988). The use or performance of bioassays is an essential part of the hazard estimation and their control procedures of the chemicals which are toxic to humans (Auletta et al., 1993 and Kirkland, 1993). Developed by B.N. Ames and co-workers in 1970's (Ames et al., 1973 and 1975) the *Salmonella* assay has been used globally for nearly 4 decades, to estimate the mutagenicity of pure chemicals and complex environmental mixtures. The test was detects a large variety of mutagens, including those activated by mammalian liver enzymes.

The reverse mutation of *Escherichia coli MTCC 40* tryptophan detects trp– to trpC reversion at a site blocking the biosynthesis of tryptophan previous to the formation of anthranilic acid. The *MTCC 40* strains of *E.coli* all carry the same AT base pair at the mutation site within the trpE gene. Nowadays the assay is used by laboratories in conjunction with the Ames *Salmonella* assay for screening chemicals for checking mutagenic activity of these chemicals. In general terms the MTCC 40 strains are used as a substitute for, or as an addition to *Salmonella strain* TA102 which also have an AT base pair at the mutation site. The assay is also combine together with the Ames assay for data submission to regulatory agencies. National and international guidelines have been investigated for performing these mutagenicity assays. The *E. coli* WP2 assay procedures are the same as those described elsewhere in this volume for the Ames *Salmonella* assay (Mortelmans and Zeiger, 2000) with the exception that limited tryptophan instead of limited histidine is used.

The *E. coli* strain MTCC 40 notice that is a radiation resistant derivative of *E. coli* B/r which is the strain used by (Luria and Delbrück 1943) in their studies that estimate that mutations occur in bacteriato tryptophan and that mutations in the other strains arise during repair of DNA damage by the 'error prone' or 'recombination' repair system. Hill reported on the extreme sensitivity of strain *e.coli* MTCC40 *uvrA* to ultraviolet light irradiation (UV), as well as its susceptibility to enhanced reversion to tryptophan independence.

There is an association between mutagenicity and carcinogenicity, based on the empirical note that the mutagens which was detected in bacterial systems have been found to be carcinogens. The Ames test has several aids over the use of mammals for the testing compounds Mutagenicity assays is cost effective, only a few days are required for testing a chemical compound and the test is performed with microgram quantities of the material. Such assays are performed on approximately 100 million organisms rather than on a limited number of animals. Thus the Ames test is used widely, which accounts for its validity and anextensive database of tested chemicals (Kier et al., 1986, Ashby and Tennant, 1988 and Ashby et al., 1991).

The Base-level testing of textile dye products requires two in vitro assays, a bacterial test (usually reverse mutations in *S. typhimurium*) and a mammalian cell genotoxicity test (MCGT). If available, in vivo results are regarded as superior to in vitro tests (Schneider et al.,2004).

In assays on more than 300 chemicals, the tryptophan requiring *E. coli* strains responded positively to 90% of the known carcinogens (Ames et al., 1973, Ashby and Tennant, 1988 and Ashby et al., 1989). The genetic damage expressed by the *Salmonella* assay represents a class of DNA damage called gene or point mutation. The*Salmonella/microsome* assay is also based on the premises that bacterial assay systems provide an efficient way to detect agents, which could interact with DNA and cause mutations

CHAPTER 4

EQUIPMENTS, MATERIALS AND EXPERIMENTALSETUP

Equipment's:

Distinctive hardware needed for the undertaking work like Incubator, Laminar Air Flow,Autoclave, Hot Air Oven, Microwave, U.V Spectrophotometer, Refrigerator utilized as a part ofthis work were benefitted from school of Biotechnology and Biosciences.

Materials:

Glass wares, Chemicals and reagents used in this work were availed by School of Biotechnology and biosciences.

Experimental Setup:

For this study, experiment was set up in Project Lab (28-404)

Methods:

Selection of Sample Site:

The dyes mention below are selected from the LUDHIANA textile mill

- Jaka red dye (vat dye)
- BROWN 3BS DYE (dispersed dye)
- TORK BLUE DYE (vat dye)
- ROYAL BLUE DYE(vat dye)
- YELLOW DYE(vat dye)
- DYCORN RED DYE (Vat dye)

Media

<u>Davis minimal agar</u>

Dextrose.. 1.0 g

Dipotassium Phosphate... 7.0 g

Monopotassium Phosphate...2.0 g

Sodium Citrate...0.5 g

Magnesium Sulfate...0.1 g

Ammonium Sulfate.. 1.0 g

Agar...15.0 g

Davis minimal agar is used for the isolation and characterization of nutritional mutants of *Escherichia coli*. Davis minimal broth (without dextrose) is used with addition dextrose in it for the isolation and characterization of nutritional mutants of *Escherichia coli* (Lederberg1 et.al)

Chapter: 5

AMES TEST

A Bruce Ames in (1928) and his students tested large numbers of commercial products in student labs at UC Berkeley when the test was first introduced in 1970s many common items, such as food colors and hairspray, were find to be mutagenic and were remove from the market. Ames also find that many mutagenic compounds are also investigated to be a **carcinogenic**, the changes in **DNA** sequence led to cancer.

The Ames test is widely to know whether the given chemical cause mutation in the DNA or not by the use of bacteria. It is a biological assay to assess the mutagenicity of the chemical. A positive test reveals that the chemical is mutagenic and may act as a carcinogenic because cancer is cause due to mutation. In the absence of a histidine/ tryptophan cell can't grow or form colonies. If there is histidine (his+)/ tryptophan (try+) present in the media only then bacteria able to grow and from colonies and when the mutagen is added in the histidine (his+)/ tryptophan plate the number of colonies increases per plate.

Experimental procedure for Ames test

1. Steps taken before to performing the experiment:

- Firstly, the *E. coli* cultures inoculated for 15–18 h to performing the experiment.
- And then Count the DAVIS agar plates and sterile plates for each plate for each chemical test.
- Prepare dilution of the chemical.
- Melt top agar with 0.05Mm tryptophan and keep at 43–48_C.:

2. When the agar was hardened (2–3 min), invert the plates are and placed in a 37 °C incubator for 48 h.

3. Then, the colonies are counted.

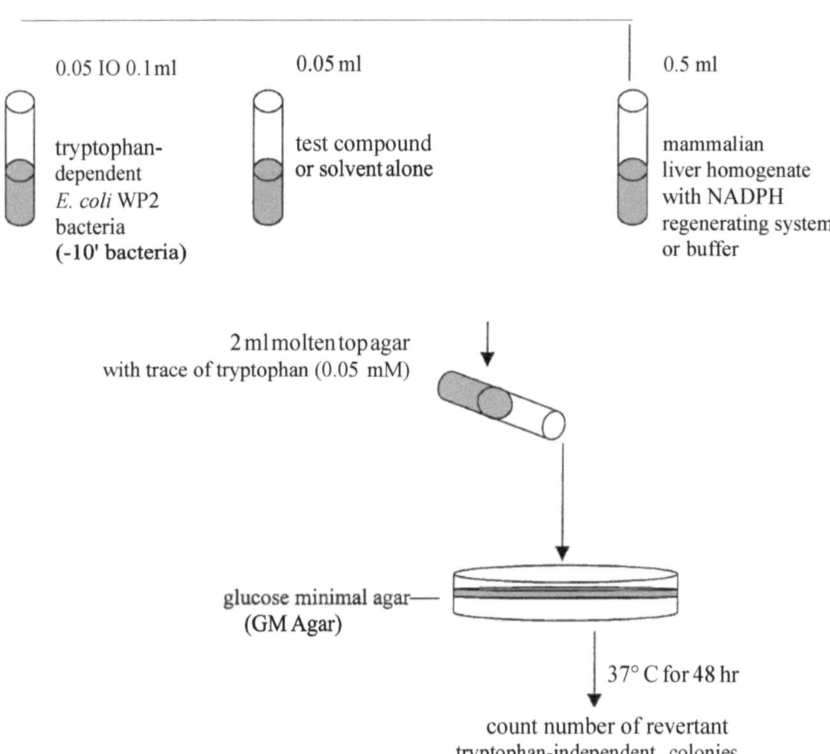
0.05 IO 0.1 ml
0.05 ml
0.5 ml
tryptophan-dependent E. coli WP2 bacteria (-10' bacteria)
test compound or solvent alone
mammalian liver homogenate with NADPH regenerating system or buffer
2 ml molten top agar with trace of tryptophan (0.05 mM)
glucose minimal agar (GM Agar)
37° C for 48 hr
count number of revertant tryptophan-independent colonies

Chapter:6

Result and discussion

To check the mutagenicity or carcenogenicity of textile dyes we take six different dyes at three different concentration i.e. at 50 ppm,150 ppm,250 ppm and perfrom two method pour plate by creating val in the media and streak plate method and observe the growth pattern of ecoli. We all know *E. coli* doesnot have the gene that code for tryptophan if we provide tryptopan externally then *E. coli* will grow. then it means sample which we taken that is dyes is only carcinogenic or cause mutation when *E. coli* grow in the media which doent have tryptophan.if not means dyes are safe to us. Let see the growth of *E. coli* at different concentration of six different types of dyes i.e. at 50 ppm.150 ppm or 250 ppm..

At 50 ppm concentration

(a) Torquoise blue dye (50 ppm)

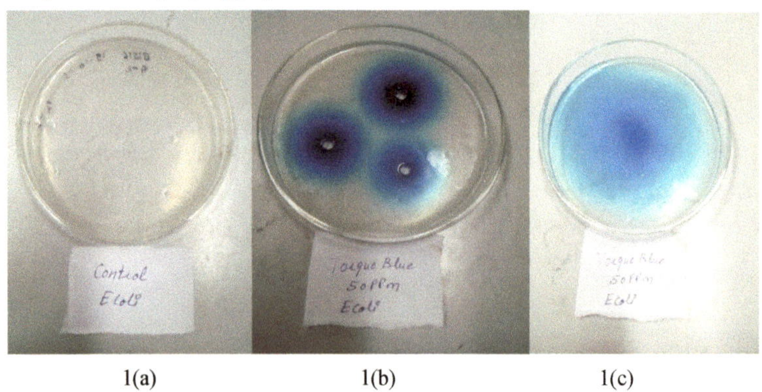

1(a) 1(b) 1(c)

Fig1(a) (control plate) having davis agar tryptophan containing media inoculated with *E. coli* bacteria in which growth was observed because of the presence mutant gene i.e. tryptophan

Fig1(b)having davis agar media in which tryptophan is not present. This plate was inoculated with *E. coli* and then by (well method) 50 ppm torquoise blue dye was added into the wells.

Fig1(c)having davis agar media in which tryptophan is not present. This plate also inoculated with *E. coli* and then by (spread plate method) 50 ppm torquoise blue dye was spreaded over it.

(a) **Yellow dye (50 ppm)**

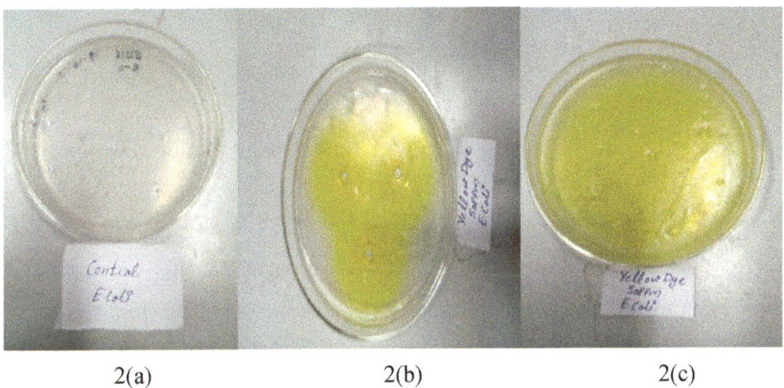

2(a) 2(b) 2(c)

Fig2(a)(control plate) having davis agar tryptophan containing media inoculated with *E. coli* bacteria in which growth was observed because of the presence mutant gene i.e. tryptophan

Fig2(b)having davis agar media in which tryptophan is not present. This plate was inoculated with *E. coli* and then by (well method) 50 ppm yellow dye was added into the wells.

Fig2(c)having davis agar media in which tryptophan is not present. This plate also inoculated with *E. coli* and then by (spread plate method) 50 ppm yellow dye was spreaded over it.

(b) Royal blue dye(50 ppm)

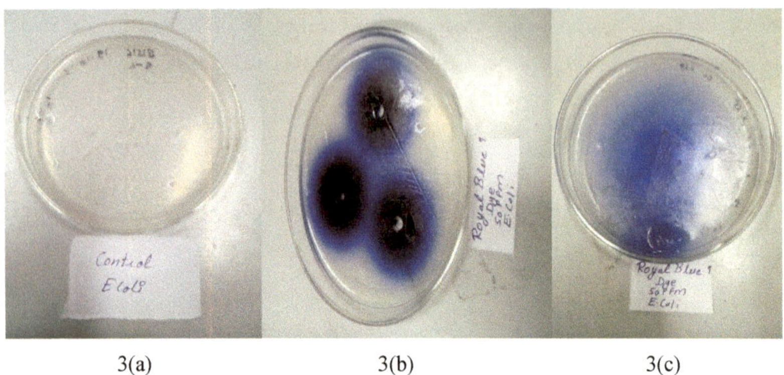

3(a) 3(b) 3(c)

Fig3(a)(control plate) having davis agar tryptophan containing media inoculated with *E. coli* bacteria in which growth was observed because of the presence mutant gene i.e. tryptophan

Fig3(b)having davis agar media in which tryptophan is not present. This plate was inoculated with *E. coli* and then by (well method) 50 ppm royal blue dye was added into the wells.

Fig3(c)having davis agar media in which tryptophan is not present. This plate also inoculated with *E. coli* and then by (spread plate method) 50 ppm royal blue dye was spreaded over it.

(c) Brown 3BS dye (50 ppm)

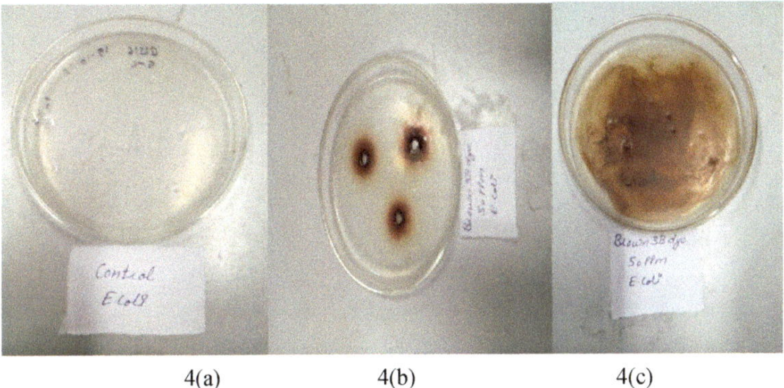

4(a) 4(b) 4(c)

Fig4(a)(control plate) having davis agar tryptophan containing media inoculated with *E. coli* bacteria in which growth was observed because of the presence mutant gene i.e. tryptophan

Fig4(b)having davis agar media in which tryptophan is not present. This plate was inoculated with *E. coli* and then by (well method) 50 ppm brown 3BS dye was added into the wells.

Fig4(c)having davis agar media in which tryptophan is not present. This plate also inoculated with *E. coli* and then by (spread plate method) 50 ppm brown 3BS dye was spreaded over it.

(d) Dycoron red dye(50 ppm)

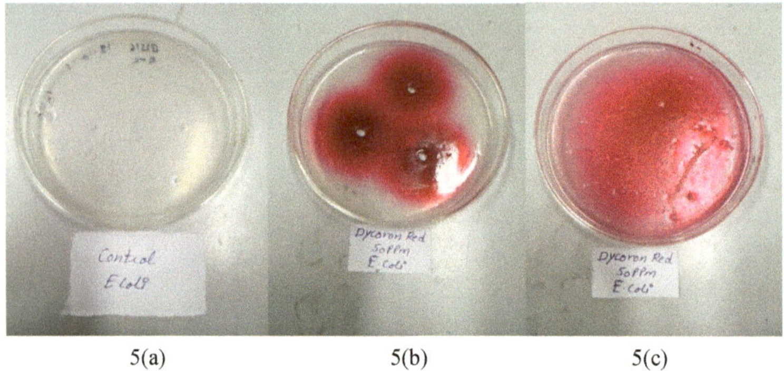

5(a) 5(b) 5(c)

Fig5(a)(control plate) having davis agar tryptophan containing media inoculated with *E. coli* bacteria in which growth was observed because of the presence mutant gene i.e. tryptophan

Fig5(b)having davis agar media in which tryptophan is not present. This plate was inoculated with *E. coli* and then by (well method) 50 ppm dycoron dye was added into the wells.

Fig5(c)having davis agar media in which tryptophan is not present. This plate also inoculated with *E. coli* and then by (spread plate method) 50 ppm dycoron dye was spreaded over it.

(e) Jaka red dye (50 ppm)

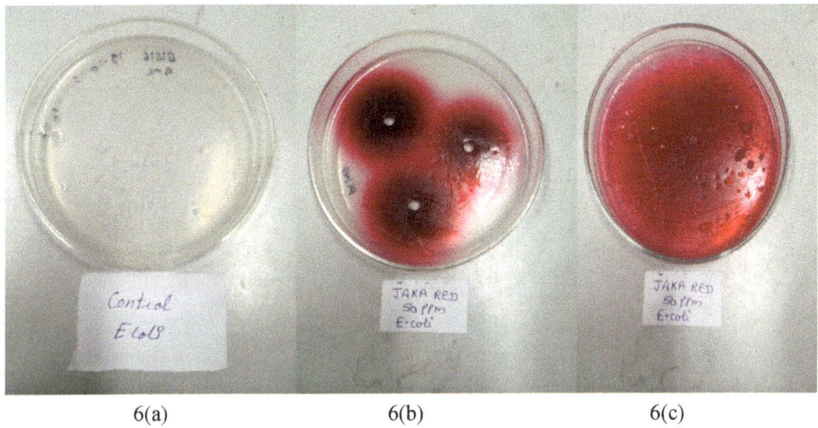

6(a) 6(b) 6(c)

Fig6(a)(control plate) having davis agar tryptophan containing media inoculated with *E. coli* bacteria in which growth was observed because of the presence mutant gene i.e. tryptophan

Fig6(b)having davis agar media in which tryptophan is not present. This plate was inoculated with *E. coli* and then by (well method) 50 ppm jaka red dye was added into the wells.

Fig6(c)having davis agar media in which tryptophan is not present. This plate also inoculated with *E. coli* and then by (spread plate method) 50 ppm jaka red dye was spreaded over it.

At 150 ppm concentration

(a) Brown 3BS dye(150 ppm)

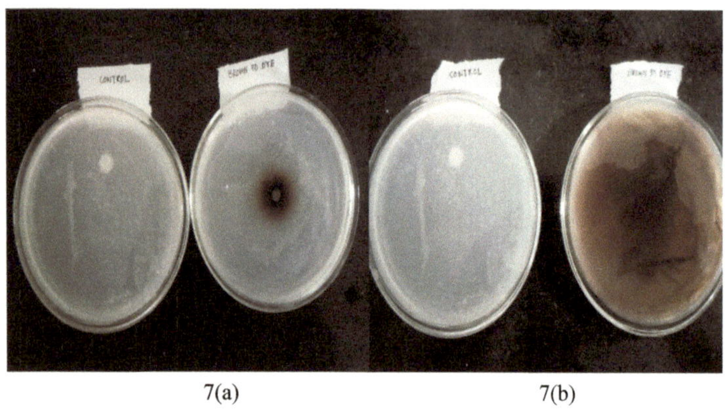

7(a) 7(b)

Fig7(a)(control plate) having davis agar tryptophan containing media inoculated with *E. coli* bacteria in which growth was observed because of the presence mutant gene i.e. tryptophan and a plate having davis agar media in which tryptophan is not present. This plate was inoculated with *E. coli* and then by (well method) 150 ppm Brown 3BS dye was added into the wells.

Fig7(b)(control plate) having davis agar tryptophan containing media inoculated with *E. coli* bacteria in which growth was observed because of the presence mutant gene i.e. tryptophan,and having davis agar media in which tryptophan is not present. This plate also inoculated with *E. coli* and then by (spread plate method) 150 ppm Brown 3BS was spreaded over it.

(a) Torquoise blue (150 ppm)

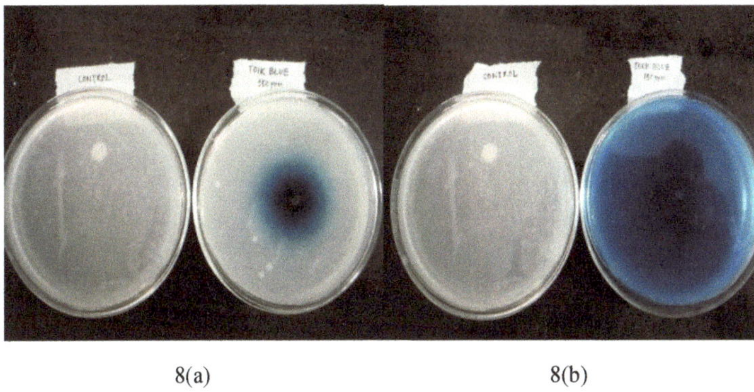

8(a) 8(b)

Fig8(a)(control plate) having davis agar tryptophan containing media inoculated with *E. coli* bacteria in which growth was observed because of the presence mutant gene i.e. tryptophan and a plate having davis agar media in which tryptophan is not present.This plate was inoculated with *E. coli* and then by (well method) 150 ppm torquoise blue dye was added into the wells.

Fig8(b)(control plate) having davis agar tryptophan containing media inoculated with *E. coli* bacteria in which growth was observed because of the presence mutant gene i.e. tryptophan,and having davis agar media in which tryptophan is not present. This plate also inoculated with *E. coli* and then by (spread plate method) 150 ppm torquoise blue dye was spreaded over it.

(b) Dycoron red dye

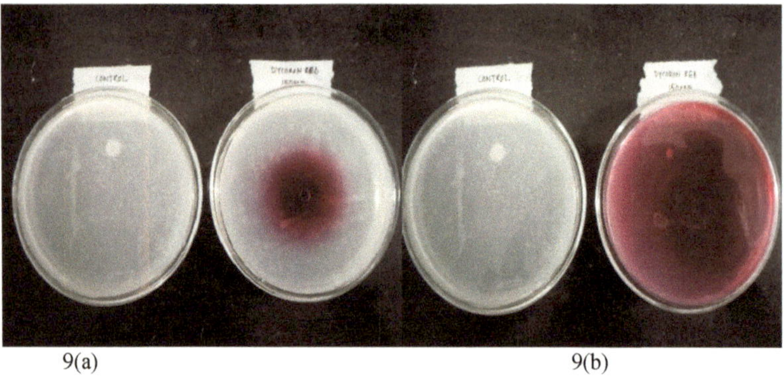

9(a) 9(b)

Fig9(a)(control plate) having davis agar tryptophan containing media inoculated with *E. coli* bacteria in which growth was observed because of the presence mutant gene i.e. tryptophan and a plate having davis agar media in which tryptophan is not present. This plate was inoculated with *E. coli* and then by (well method) 150 ppm Dycoron red dye was added into the wells.

Fig9(b)(control plate) having davis agar tryptophan containing media inoculated with *E. coli* bacteria in which growth was observed because of the presence mutant gene i.e. tryptophan,and having davis agar media in which tryptophan is not present. This plate also inoculated with *E. coli* and then by (spread plate method) 150 ppm Dycoron red dye was spreaded over it.

(c) Royal blue (150 ppm)

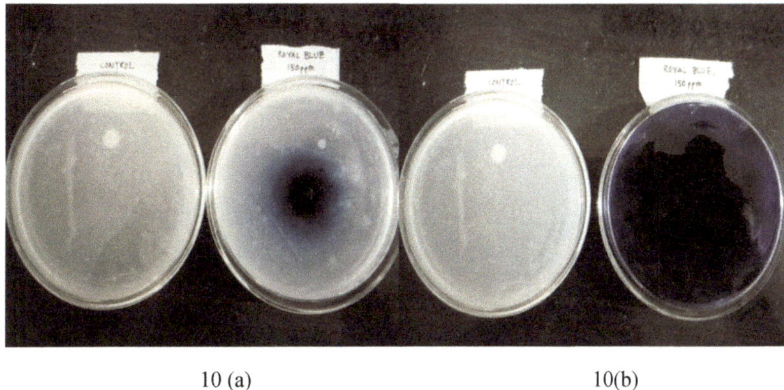

10 (a) 10(b)

Fig10(a)(control plate) having davis agar tryptophan containing media inoculated with *E. coli* bacteria in which growth was observed because of the presence mutant gene i.e. tryptophan and a plate having davis agar media in which tryptophan is not present. This plate was inoculated with *E. coli* and then by (well method) 150 ppm royal blue dye was added into the wells.

Fig10(b)(control plate) having davis agar tryptophan containing media inoculated with *E. coli* bacteria in which growth was observed because of the presence mutant gene i.e. tryptophan,and having davis agar media in which tryptophan is not present. This plate also inoculated with *E. coli* and then by (spread plate method) 150 ppm royal blue dye was spreaded over it.

(d) Yellow dye (150 ppm)

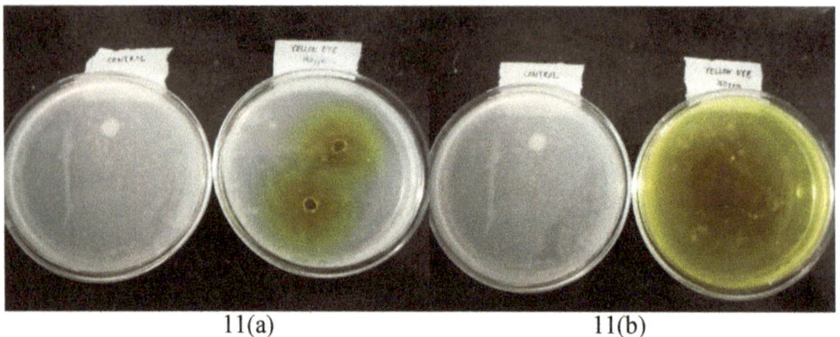

11(a) 11(b)

Fig11(a)(control plate) having davis agar tryptophan containing media inoculated with *E. coli* bacteria in which growth was observed because of the presence mutant gene i.e. tryptophan and a plate having davis agar media in which tryptophan is not present. This plate was inoculated with *E. coli* and then by (well method) 150 ppm yellow red dye was added into the wells.

Fig11(b)(control plate) having davis agar tryptophan containing media inoculated with *E. coli* bacteria in which growth was observed because of the presence mutant gene i.e. tryptophan,and having davis agar media in which tryptophan is not present. This plate also inoculated with *E. coli* and then by (spread plate method) 150 ppm yellow red dye was spreaded over it.

(e) Jaka red (150 ppm)

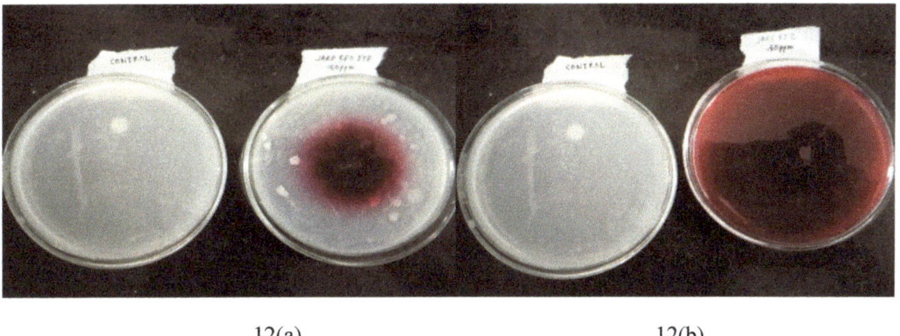

12(a) 12(b)

Fig12(a)(control plate) having davis agar tryptophan containing media inoculated with *E. coli* bacteria in which growth was observed because of the presence mutant gene i.e. tryptophan and a plate having davis agar media in which tryptophan is not present. This plate was inoculated with *E. coli* and then by (well method) 150 ppm jaka red dye was added into the wells.

Fig12(b)(control plate) having davis agar tryptophan containing media inoculated with *E. coli* bacteria in which growth was observed because of the presence mutant gene i.e. tryptophan,and having davis agar media in which tryptophan is not present. This plate also inoculated with *E. coli* and then by (spread plate method) 150 ppm jaka red dye was spreaded over it.

At 250 ppm concentration

(a) Yellow dye (250 ppm)

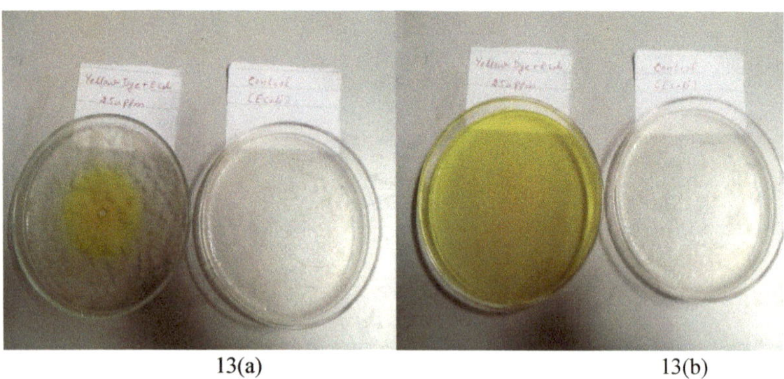

13(a) 13(b)

Fig13(a)(control plate) having davis agar tryptophan containing media inoculated with *E. coli* bacteria in which growth was observed because of the presence mutant gene i.e. tryptophan and a plate having davis agar media in which tryptophan is not present. This plate was inoculatedwith *E. coli* and then by (well method) 250 ppm yellow dye was added into the wells.

Fig13(b)(control plate) having davis agar tryptophan containing media inoculated with *E. coli* bacteria in which growth was observed because of the presence mutant gene i.e. tryptophan,and having davis agar media in which tryptophan is not present. This plate also inoculated with *E. coli* and then by (spread plate method) 250 ppm yellow dye was spreaded over it.

(b) Jaka red dye (250 ppm)

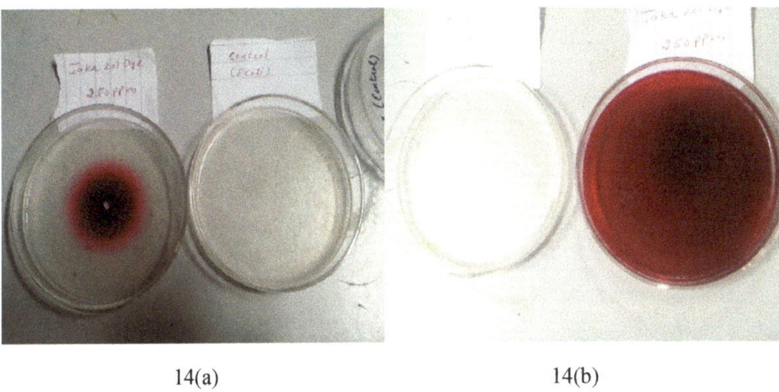

14(a) 14(b)

Fig14(a)(control plate) having davis agar tryptophan containing media inoculated with *E. coli* bacteria in which growth was observed because of the presence mutant gene i.e. tryptophan and a plate having davis agar media in which tryptophan is not present. This plate was inoculated with *E. coli* and then by (well method) 250 ppm jaka 201 dye was added into the wells.

Fig14(b)(control plate) having davis agar tryptophan containing media inoculated with *E. coli* bacteria in which growth was observed because of the presence mutant gene i.e. tryptophan,and having davis agar media in which tryptophan is not present. This plate also inoculated with *E. coli* and then by (spread plate method) 250 ppm jaka 201 dye was spreaded over it.

(c) Dycoron red(250 ppm)

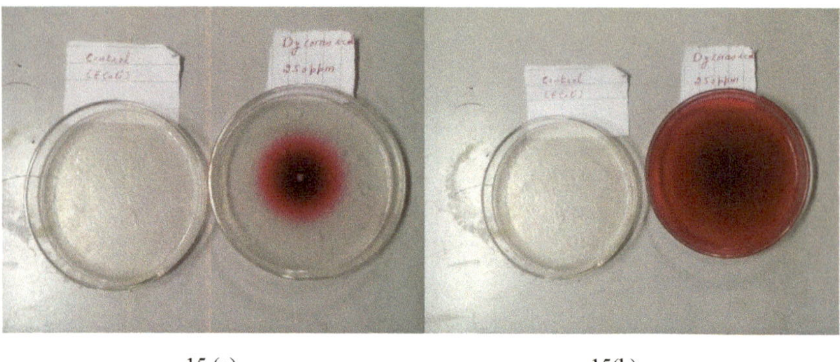

15 (a) 15(b)

Fig15(a)(control plate) having davis agar tryptophan containing media inoculated with *E. coli* bacteria in which growth was observed because of the presence mutant gene i.e. tryptophan and a plate having davis agar media in which tryptophan is not present. This plate was inoculated with *E. coli* and then by (well method) 250 ppm dycoron red dye was added into the wells.

Fig15(b)(control plate) having davis agar tryptophan containing media inoculated with *E. coli* bacteria in which growth was observed because of the presence mutant gene i.e. tryptophan,and having davis agar media in which tryptophan is not present. This plate also inoculated with *E. coli* and then by (spread plate method) 250 ppm dycoron red dye was spreaded over it.

(d) Brown 3BS dye (250 ppm)

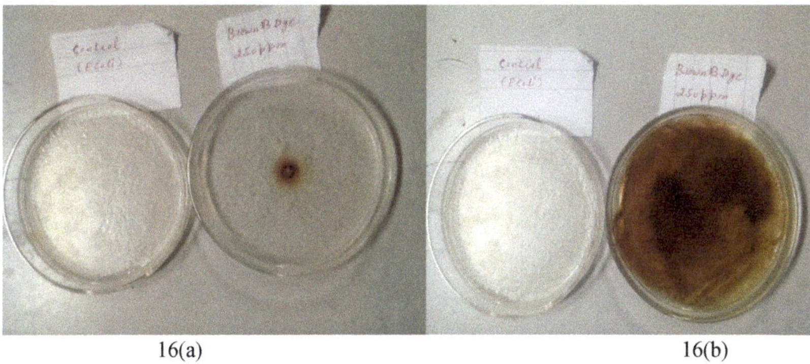

16(a) 16(b)

Fig16(a)(control plate) having davis agar tryptophan containing media inoculated with *E. coli* bacteria in which growth was observed because of the presence mutant gene i.e. tryptophan and a plate having davis agar media in which tryptophan is not present. This plate was inoculated with *E. coli* and then by (well method) 250 ppm brown 3BS dye was added into the wells.

Fig16(b)(control plate) having davis agar tryptophan containing media inoculated with *E. coli* bacteria in which growth was observed because of the presence mutant gene i.e. tryptophan,and having davis agar media in which tryptophan is not present. This plate also inoculated with *E. coli* and then by (spread plate method) 250 ppm brown 3BS dye was spreaded over it.

(e) Torquoise blue(250 ppm)

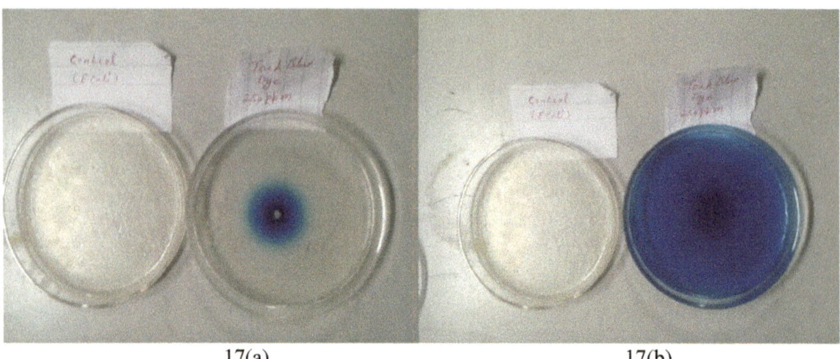

17(a) 17(b)

Fig17(a)(control plate) having davis agar tryptophan containing media inoculated with *E. coli* bacteria in which growth was observed because of the presence mutant gene i.e. tryptophan and a plate having davis agar media in which tryptophan is not present. This plate was inoculated with *E. coli* and then by (well method) 250 ppm Torquoise blue dye was added into the wells.

Fig17(b)(control plate) having davis agar tryptophan containing media inoculated with *E. coli* bacteria in which growth was observed because of the presence mutant gene i.e. tryptophan,and having davis agar media in which tryptophan is not present. This plate also inoculated with *E. coli* and then by (spread plate method) 250 ppm Torquoise blue dye was spreaded over it.

(f) Royal blue dye (250 ppm)

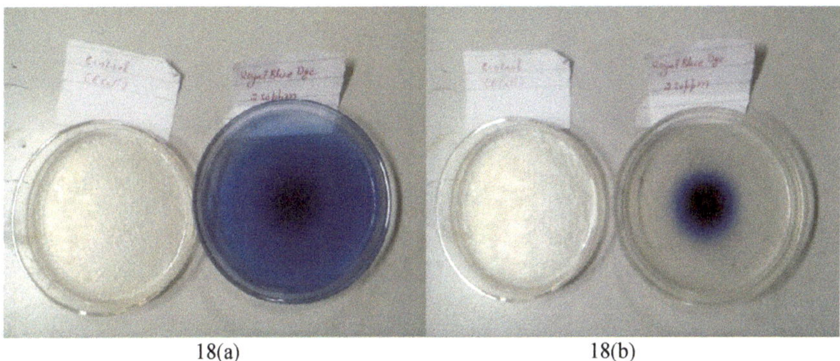

18(a) 18(b)

Fig18(a)(control plate) having davis agar tryptophan containing media inoculated with *E. coli* bacteria in which growth was observed because of the presence mutant gene i.e. tryptophan,and having davis agar media in which tryptophan is not present. This plate also inoculated with *E. coli* and then by (spread plate method) 250 ppm royal blue dye was spreaded over it.

Fig18(b)(control plate) having davis agar tryptophan containing media inoculated with *E. coli* bacteria in which growth was observed because of the presence mutant gene i.e. tryptophan and a plate having davis agar media in which tryptophan is not present. This plate was inoculated with *E. coli* and then by (well method) 250 ppm royal blue dye was added into the wells.

Data analysis

As we see in all concentration in the absence of tryptophan growth takes places it means that all dyes are carcinogenic in nature. As shown above three concentrations was taken 50 ppm, 150 ppm, 250 ppm which shown that at what amount dyes show there more effect. In this we predict that as we increase the amount of dyes they are more dangerous to the peoples living close to the area. This test specifies that if a test compounds doubles or more than doubles, that mean spontaneous mutation is occured and if there is no grow means no mutation in the test compound.

Result is positive

As we see to check the mutagenicity of textile dispersed or vat dyes ames test was used in which different concentration of the dyes where taken i.e.50ppm,150ppm and 250ppm. The dyes which we used are royal blue dye, brown 3 BS dye, yellow dye, Torquoise blue dye, dycoron red dye and also one e.coli strain i.e MTCC 40 strain was taken which lack tryptophan mutant gene.

Davis minimal agar is used for the isolation and characterization nutritional mutants of Escherichia coli. Then we culture it by using two methods i.e. by well method or by spread plate method as shown above in figures 1 to 18. Firstly 50ppm concentration of the dyes was taken where we observe growth on control because in control we add tryptophan to the media and e.coli lack tryptophan mutant gene it obvious that e.coli will grow. The rest two plate also showing growth but in less number from observing growth in the media it was clear that the dyes which we use in textile industries as a coluring agent are carcinogenic and cause mutation when they enter in our body through water or any other route. Then after that we want to know that how much concentration these dyes are more carcinogenic then we take two different concentration of the dyes i.e. 150ppm 250ppm and use same procedure for that which are used for 50ppm concentration and compare all the three concentration. At last we found that as the amount of dyes dispose in water gets increase day by day it will more harmful for the people living nereby area and it is more carcinogenic because at 50ppm we observe less growth of e.coli as we increase the concentration the increases which shown that as the concentration of the dyes increases it is dangerous for the peoples and more step should be taken for the welfare of public health.

Chapter: 7

Conclusion

The present investigation during the period from August-May 2016 at the department of Microbiology, Lovely Professional University, Phagwara (Punjab).the specific objectives were to Survey of Ludhiana textile mills for knowing different disperse and vat dyes used throughout the year and To perform an assay to determine the mutagenicity of disperse and vat dyes (AMES test). Then we survey Ludhiana and knowing about different vat or dispersed dyes i.e. royal blue dye, Jaka red DYE, dycoron red dye, yellow dyes etc. which was used recently to coloring the clothes and this enter into the nearby area when they dispose it out. After that we take these dyes sample and perform Ames test and found that all dyes was carcinogenic in nature and cause mutation when enter in our body. Therefore, if further studies and research are done, the consequences for the environment can be fruitful

Chapter: 8 REFERENCES

1) Clasen, T., Schmidt, W., Rabie, T., Roberts, I., & Cairncross, S. (2007). Interventions to improve waterquality for preventing diarrhoea: Systematic review and meta-analysis, *British Medical Journal.* Cited November 22, 2009 http://www.bmj.com/cgi/reprint/334/7597/782.

2) Anderson, B. A., Romani, J. H., Phillips, H. E., & van Zyl, J. A. (2002). Environment, access to health care, and other factors affecting infant and child survival among the African and Coloured populations of South Africa, 1989-94. *Population and Environment, 23*, 349-364.Vidyasagar, D. (2007). Global minute: water and health - walking for water and water wars *Journal of Perinatology* 27, 56–58.

3) Ross, J. A., Rich, M., Molzen, J., and Pensak, M.. 1988. *Family planning and child survival in onehundred developing countries.* New York: Center for Population and Family Health, Columbia University.

4) Fewtrell, Lorna, Kaufmann, Rachel B., Kay, David, Enanoria, Wayne, Haller, Laurence, &Colford, John M., Jr. 2005."Water, sanitation, and hygiene interventions to reduce diarrhoea in less developed countries: a systematic review and meta-analysis," *The Lancet Infectious Diseases*, 5: 42-52.

5) Laurent, P. 2005. Household drinking water systems and their impact on people with weakened immunity,MSF-Holland, Public Health Department, February. Cited June 7, 2007._ http://www.who.int/household_water/research/HWTS_impacts_on_weakened_immunity.pdf

6) Kgalushi, R., Smite, S., &Eales, K. (2004). "People living with HIV/AIDS in a context of rural poverty:the importance of water and sanitation services and hygiene education," Johannesburg: Mvula Trustand Delft: IRC International Water and Sanitation Centre. Cited June 23, 2008. http://www.irc.nl/page/10382

7) Rafi, F., Franklin, W. and Cerniglia, C.E. (1990). azoreductase activity ofanaerobic bacteria isolated from human intestinal microflora. Apply Environ. Microbiol., 56:2146-2151.

8) ETAD (1997). German ban of use of certain azo compounds in someconsumer goods. ETAD Information Notice No. 6, (Revised).

9) Kirkland, D. J. (1993): Genetic toxicology testingrequirements: Official and unofficial views fromEurope. Environmental and MolecularMutagenesis, 21: 8-14

10) Waters, M. D., Stack, H. F., Brady, A. L., Lohman, P. H. M., Haroun, S. and Vainio (1988): Use of computerized data listings and activity profiles of genetic and related effects in the review of 195 compounds. *Mut. Res.*, 205: 295-312.

11) Auletta, A. E., Dearfield, K. L. and Cimino, M. C. (1993): Mutagenicity Test Schemes and Guidelines: USEPA Office of Pollution Prevention and Toxics and Office of Pesticide Programs.*Environmental and Molecular Mutagensis,* 21:38-45.

12) Ames, B. N., Durston, W. E., Yamasaki, E. and Lee, F. D. (1973): Carcinogens are mutagens: a simple test system combining liverhomogenates for activation and bacteria for detection. *Proc. of the Nat. Acad. of Sci. U.S.A.,*70: 2281-2285.

13) D. Gatehouse, S. Haworth, T. Cebula, E. Gocke, L. Kier, T. Matsushima, C. Melcion, T. Nohmi, T. Ohta, S. Venitt, E. Zeiger, Recommendations for the performance of bacterial mutation assays, Mutat. Res. 312 (1994) 217–233.

14) S.E. Luria, M. Delbrück, Mutations in bacteria from virus sensitivity to virus resistance, Genetics 28 (1943) 491–511.

15) K. Mortelmans, B.A.D. Stocker, Segretation of the mutatorproperty of plasmid R46 from its ultraviolet-protectionproperties, Mol Gen. Genet. 167 (1979) 317–327.

16) Kirkland, D. J. (1993): Genetic toxicology testingrequirements: Official and unofficial views fromEurope. Environmental and MolecularMutagenesis, 21: 8-14

17) Ashby, J. and Tennant, R. W. (1988): Chemical structure, Salmonella mutagenicity and extentfrom a Dyeing Factory in Hong Kong. Environ.Toxicol., 18: 312–316.

18) Schneider, K., Hafner, C. and Jager, I. (2004):Mutagenity of Textile Dye Products. J. Appl.Toxicol., 24: 83–91.

19) 1Lederberg. 1950. Methods in Med. Res. 3:5.

20) Davis. 1949. Proc. Natl. Acad. Sci. 35:1.

21) Nester, Schafer and Lederberg. 1963. Genetics 48:529.

YOUR KNOWLEDGE HAS VALUE

- We will publish your bachelor's and master's thesis, essays and papers
- Your own eBook and book - sold worldwide in all relevant shops
- Earn money with each sale

Upload your text at www.GRIN.com and publish for free